BoV

Spinosaurus

Lori Dittmer

CREATIVE EDUCATION
CREATIVE PAPERBACKS

seedlings

Published by Creative Education and Creative Paperbacks
P.O. Box 227, Mankato, Minnesota 56002
Creative Education and Creative Paperbacks
are imprints of The Creative Company
www.thecreativecompany.us

Design by Ellen Huber
Production by Rachel Klimpel and Ciara Beitlich
Art direction by Rita Marshall

Photographs by 123RF (Orla), Getty (Dorling Kindersley), iStock (Elenarts108, jondpatton, Ieonello), Science Source (Animate4.com, Animate4.com/Science Photo Library, CARLTON PUBLISHING GROUP, Mark Garlick/Science Photo Library, MASATO HATTORI), Shutterstock (Elle Arden Images, Herschel Hoffmeyer, Matis75), ThinkstockPhotos (azutura, LindaMarieB)

Copyright © 2024 Creative Education, Creative Paperbacks
International copyright reserved in all countries.
No part of this book may be reproduced in any form
without written permission from the publisher.

Library of Congress Cataloging-in-Publication Data
Names: Dittmer, Lori, author.
Title: Spinosaurus / by Lori Dittmer.
Description: Mankato, Minnesota : Creative Education and Creative Paperbacks, [2024] | Series: Seedlings: dinosaurs | Includes bibliographical references and index. | Audience: Ages 4–7 | Audience: Grades K–1 | Summary: "Early readers are introduced to Spinosaurus, a Cretaceous carnivore with a large sail. Friendly text and dynamic photos share the dinosaur's looks, behaviors, and diet, based on scientific research"— Provided by publisher.
Identifiers: LCCN 2022015647 (print) | LCCN 2022015648 (ebook) | ISBN 9781640265035 (library binding) | ISBN 9781682770559 (paperback) | ISBN 9781640006331 (ebook)
Subjects: LCSH: Spinosaurus—Juvenile literature. | Dinosaurs—Juvenile literature.
Classification: LCC QE862.S3 D586 2024 (print) | LCC QE862.S3 (ebook) | DDC 567.912—dc23/eng/20221027
LC record available at https://lccn.loc.gov/2022015647
LC ebook record available at https://lccn.loc.gov/2022015648

Printed in China

TABLE OF CONTENTS

Hello, *Spinosaurus!* 4

Cretaceous Dinosaurs 6

Early Discovery 8

Spines and Sails 10

Swimming Giants 12

Meat Eaters 14

What Did *Spinosaurus* Do? 16

Goodbye, *Spinosaurus!* 18

Picture a *Spinosaurus* 20

Words to Know 22

Read More 23

Websites 23

Index 24

Hello, *Spinosaurus!*

This dinosaur lived long ago.

Iguanodon and *Triceratops* lived at the same time.

We know of *Spinosaurus* from its fossils. It was first found in Egypt in 1912.

Spinosaurus means "spine lizard." Its spine came out of its back.

The bones formed a sail as tall as an adult human!

Huge *Spinosaurus* was longer than a school bus! It moved on two legs.

It probably swam in rivers, too. Its tail was shaped like a paddle.

Spinosaurus ate meat, including fish.

It trapped food in its long, narrow snout.

Spinosaurus walked on land.

It went into the water.

It hunted for food.

Goodbye, *Spinosaurus*!

Picture a *Spinosaurus*

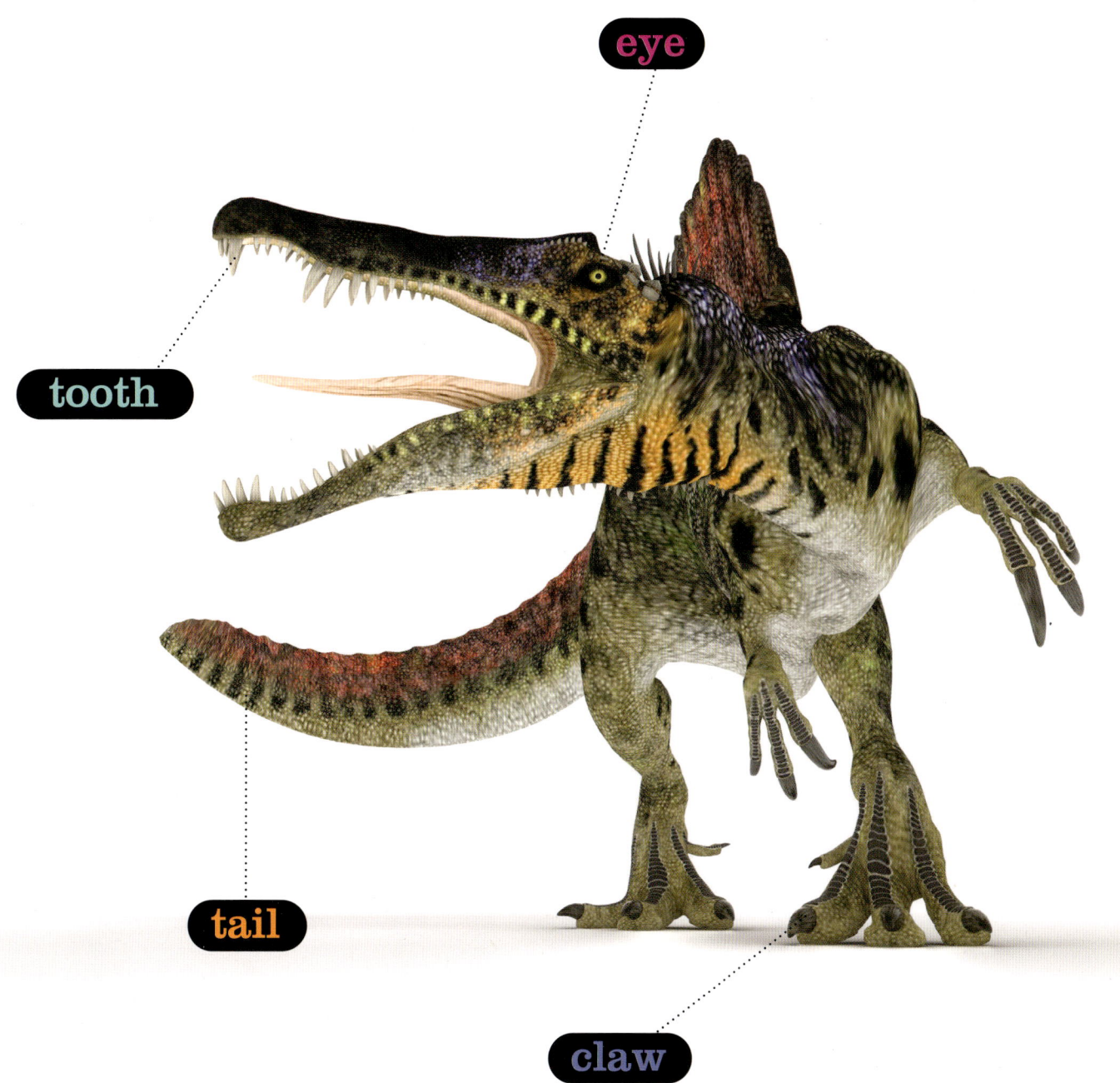

Words to Know

fossil: a bone or trace from an animal long ago that can be found in some rocks

paddle: a short pole with a wide, flat part at the end

snout: the part of an animal's face that sticks out and includes the nose and mouth

spine: a large spike of bone

Read More

Nelson, Jake. *I'm a Spinosaurus*. Ann Arbor, Mich.: Cherry Lake Publishing, 2021.

Pimentel, Annette Bay. *Do You Really Want to Meet Spinosaurus?* Mankato, Minn.: Amicus, 2020.

Websites

DK Find Out! | *Spinosaurus*
https://www.dkfindout.com/us/dinosaurs-and-prehistoric-life/dinosaurs/spinosaurus
Read more about *Spinosaurus* and take a dinosaur quiz.

National Geographic Kids | *Spinosaurus*
https://kids.nationalgeographic.com/animals/prehistoric/facts/spinosaurus
Learn more about *Spinosaurus* and watch a video.

Note: Every effort has been made to ensure that the websites listed above are suitable for children, that they have educational value, and that they contain no inappropriate material. However, because of the nature of the Internet, it is impossible to guarantee that these sites will remain active indefinitely or that their contents will not be altered.

Index

discovery, 8
feeding, 14, 15, 17
fossils, 8
length, 12
name, 10
snout, 15
spine, 10, 11
swimming, 13
tail, 13
when it lived, 6, 7